WHS

AGE

10-11

Key Stage 2

national TESTS
practice papers

FOR THE YEAR 2002

Key Stage 2

Mental Arithmetic Book 1

practice Papers

Contents

First published 2001
exclusively for WHSmith by

Hodder & Stoughton Educational,
a division of Hodder Headline Ltd
338 Euston Road
London NW1 3BH

A CIP record for this book is available from the British Library.

Authors: Steve Mills and Hilary Koll

ISBN 0340 84603 8

Printed and bound by Graphycems, Spain

NOTE: The tests, questions and advice in this book are not reproductions of the official test materials sent to schools. The official testing process is supported by guidance and training for teachers in setting and marking tests and interpreting the results. The results achieved in the tests in this book may not be the same as are achieved in the official tests.

The National Tests: A Summary

What are the National Tests?

Children who attend state schools in England and Wales sit National Tests (SATs) at the ages of 7, 11 and 14, usually at the beginning of May. They may also sit optional tests in the intervening years – many schools have chosen to adopt these tests. The test results are accompanied by an assessment by the child's teacher.

The results are used by the school to assess each child's level of knowledge and progress in English and Maths at Key Stage 1 and English, Maths and Science at Key Stages 2 and 3. They also provide useful guidance for the child's next teacher.

The educational calendar for children aged 5-14 is structured as follows:

Key Stage	Year	Age by end of year	National Test
1 (KS1)	1	6	
	2	7	KEY STAGE 1
2 (KS2)	3	8	Optional Year 3
	4	9	Optional Year 4
	5	10	Optional Year 5
	6	11	KEY STAGE 2
3 (KS3)	7	12	
	8	13	
	9	14	KEY STAGE 3

Test Timetable

The Key Stage 1 National Tests are carried out in **May**.

Key Stage 2 tests take place in one week in May. All children sit the same test at the same time. In 2002, the tests will take place in the week of **13–17 May**. Your child's school will be able to provide you with a detailed timetable.

Key Stage 3 students will sit their tests on **7–13 May.**

Levels

National average levels have been set for children's results in the National Tests. The levels are as follows:

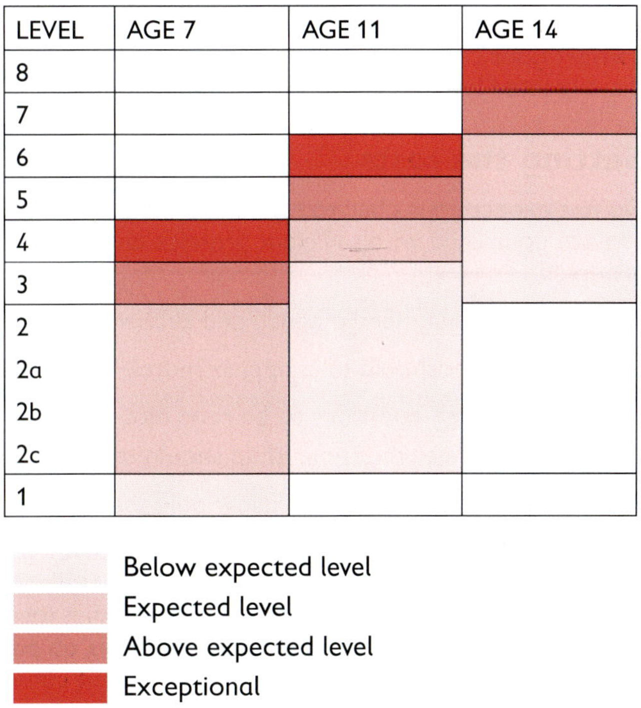

Results

Your child's school will send you a report indicating his or her levels in the tests and the teacher assessment.

The school's overall test results will be included in local and national league tables, which are published in most newspapers.

What can parents do to help?

While it is never a good idea to encourage cramming, you can help your child to succeed by:

- Making sure he or she has enough food, sleep and leisure time during the test period.
- Practising important skills such as writing and reading stories, spelling and mental arithmetic.
- Telling him or her what to expect in the test, such as important symbols and key words.
- Helping him or her to be comfortable in test conditions including working within a time limit, reading questions carefully and understanding different ways of answering.

Mental Maths at Key Stage 2

Most children at Key Stage 2 will take an orally delivered Mental Maths Test and two written tests. This book contains 12 Mental Maths Tests and practice questions for the range of topics covered.

The marks given to each child in the national Mental Maths Test are used in conjunction with the marks gained in the written tests to find out a National Curriculum level, with a typical 11-year-old attaining Level 4.

To gain an idea of the level at which your child is working, use the table on page 28, which shows you how to convert your child's marks for each Mental Maths Test into a National Curriculum level. Children working at the expected level or above should aim to score more than 11 marks out of 20 in each test.

Setting the tests

Do not expect your child to take the tests one after another. Taking the tests regularly over a few weeks can provide your child with sufficient practice and can help him or her to notice improvements and to track progress.

Your child will only need a pencil and rubber for each test.

Each mental test should take approximately 10-15 minutes to give. It consists of:

- A list of twenty questions to be read aloud – cut this out for ease of use.
- An answer sheet that your child should use for recording answers.

Read each question twice and allow only the time suggested for your child to answer.

Questions in the national Mental Maths Test cover a wider range than just calculation in your head. Where mental calculation is required, your child can make small jottings on the answer sheets to help him or her retain information from the question. These must then NOT be used to form the basis of written calculations: children are not allowed to 'work out' answers on paper. For example, if a question asks children to multiply 25 by 8, your child could jot down 25 and 8, but cannot then set it out as a written multiplication question to find the answer. Instead a mental method should be used, such as using a 'double, double, double' method for multiplying by 8, i.e. 25…50…100…200.

Marking the tests

Discuss the answers with your child and encourage him or her to explain the methods used to reach an answer. Any mental method that reaches the correct answer should be praised, although alternative and perhaps quicker methods can be discussed.

Keep a record of your child's scores for each test to monitor his or her progress.

"For this first set of questions you have five seconds to work out each answer and write it down."

1 What is seven multiplied by five?
2 How many months are there in ten years?
3 Write the number nine thousand and eighteen in figures.
4 How many grams are there in two point six kilograms?
5 What is three quarters as a decimal?
6 What number is three less than two thousand?

"For the next set of questions you have ten seconds to work out each answer and write it down."

7 A pen costs eighty pence and a pencil case costs one pound ninety nine pence. How much do the pen and pencil case cost altogether?
8 The temperature yesterday was six degrees Celsius, and the temperature today is nine degrees colder. What is the temperature today?
9 Look at the answer sheet. In the equation what is the value of *n*?
10 What is one quarter of eight point four?
11 Multiply eight by twenty five.
12 Two numbers have a difference of eighteen. One number is thirty: what could the other number be?
13 Look at the answer sheet. Circle the number that is nearest to six metres.
14 Molly slept for nine and a half hours without waking. She went to sleep at nine p.m. What time did she wake?
15 Look at the answer sheet. What is the mean of the three test scores?

"For the next set of questions you have fifteen seconds to work out each answer and write it down."

16 Look at the answer sheet. Draw a ring around the largest of the five numbers.
17 A regular shape has a perimeter of thirty six centimetres. Each side is six centimetres. What is the name of the shape?
18 I have sixty pence. My friend has three times that much. How much do we have altogether?
19 Which number is exactly half way between one thousand five hundred and two thousand?
20 Look at the answer sheet. What is the size of angle B?

Test 2

Questions

"For this first set of questions you have five seconds to work out each answer and write it down."

1 Write the number forty thousand and eleven in figures.
2 What is sixteen more than twenty three?
3 What is thirty two divided by ten?
4 What is two thirds of eighteen?
5 How many sides do five pentagons have in total?
6 How many hours are there in three days?

"For the next set of questions you have ten seconds to work out each answer and write it down."

7 What is half of eight point two?
8 Add six and eighteen and then divide by four.
9 A snail crawls seventy two centimetres. How much further does it have to crawl to reach one metre?
10 A cartoon programme lasts for twenty five minutes. It starts at ten minutes to seven. What time does it end?
11 A shop has fifty per cent off all prices. What is the new price of a book that was seventeen pounds?
12 On the answer sheet is part of a scale. What number is the arrow pointing to?
13 Imagine a square based pyramid. How many of its faces are triangles?
14 A one litre jug is three quarters full. How many more millilitres are needed to fill it?
15 A football team scored five goals for every two they let in. They scored twenty goals. How many did they let in?

"For the next set of questions you have fifteen seconds to work out each answer and write it down."

16 Which number is exactly half way between twenty two and forty eight?
17 Find the total of twenty eight, thirty one and forty two.
18 A man was born in 1873. What year was the man's ninety ninth birthday?
19 Each of the sides of a pentagon is five point two centimetres. What is its perimeter?
20 Look at the answer sheet. How many minutes does it take to get from the shopping centre to the cinema?

"For this first set of questions you have five seconds to work out each answer and write it down."

1 Subtract sixteen from fifty.

2 Two numbers have a total of one thousand. One of the numbers is six hundred and fifty. What is the other number?

3 Write the number that is 5 squared.

4 The time is quarter to two in the afternoon. Write this time as it would be shown on a twenty-four hour digital clock.

5 How many two pence pieces have the same value as one pound?

6 The perimeter of a square is eighty centimetres. What is the length of one of its sides?

"For the next set of questions you have ten seconds to work out each answer and write it down."

7 Find the difference between two hundred and thirty and five hundred and ten.

8 Look at the number sequence on your answer sheet. What is the next number?

9 How much change from five pounds would you get if you spent exactly seventy five pence?

10 If I add fourteen to a number I get twenty seven. What is the number?

11 A TV programme starts at twenty five to six and lasts for thirty five minutes. What time does the programme end?

12 On the answer sheet is part of a scale. What number is the arrow pointing to?

13 How many teams of seven people can be made from a group of fifty six people?

14 London is one hundred and eighty kilometres from Bristol. If I go from London to Bristol and back, how many kilometres have I travelled?

15 Circle the fraction that is equivalent to nought point three.

"For the next set of questions you have fifteen seconds to work out each answer and write it down."

16 Twenty five per cent of a number is twelve. What is the number?

17 Which number is exactly half way between sixteen and thirty?

18 A woman was born in 1913. What year was the woman's forty ninth birthday?

19 Circle the expression shown on your answer sheet that has an answer nearest to fifty.

20 Look at the pie chart on the answer sheet. One hundred and sixty people were asked what their favourite food was. How many people said their favourite food was chocolate?

Test 4

Questions

"For this first set of questions you have five seconds to work out each answer and write it down."

1 What is nine multiplied by seven?
2 How many days are there in eleven weeks?
3 Write the number ten thousand and four in figures.
4 How many grams are there in a quarter of a kilogram?
5 What is nought point two five as a fraction?
6 What is sixty multiplied by twenty?

"For the next set of questions you have ten seconds to work out each answer and write it down."

7 How many seconds are there in ten minutes?
8 Add twelve to thirteen and then multiply by four.
9 Look at the answer sheet. In the equation, what is the value of *n*?
10 One third of a number is twelve. What is the number?
11 Divide one thousand by four.
12 The time is six forty in the evening. Write this as it would be shown on a twenty-four hour digital clock.
13 Look at the answer sheet. Draw a ring around the decimal that is equivalent to thirty per cent.
14 Peter's train journey started at quarter past eight in the morning. The journey lasted one and a half hours. What time did the journey end?
15 Look at the answer sheet. What is the mode of these numbers?

"For the next set of questions you have fifteen seconds to work out each answer and write it down."

16 Look at the answer sheet. Draw a ring around the number nearest to six.
17 A rectangle has sides of three point six centimetres and six point four centimetres. What is its perimeter?
18 Imagine a sphere. If a straight cut was made through the sphere to form two identical shapes, what would these shapes be called?
19 Look at the answer sheet. Draw a ring around any of the numbers that are square numbers.
20 Look at the answer sheet. It takes seventeen minutes to get from the shopping centre to the cinema. Write what time the bus arrives at the cinema.

"For this first set of questions you have five seconds to work out each answer and write it down."

1 Write the number thirty thousand and sixteen in figures.
2 What is eleven more than twenty six?
3 Subtract sixty from two hundred and ten.
4 What is one third of twenty seven?
5 How many 5p coins have the same value as one pound?
6 How many metres are there in one and a half kilometres?

"For the next set of questions you have ten seconds to work out each answer and write it down."

7 What is half of one thousand and eighty?
8 Add six point two and three point nine.
9 A girl buys a seventy pence magazine. She pays with a ten pound note. How much change is she given?
10 A TV programme lasts for thirty five minutes. It starts at quarter past seven. What time does it end?
11 A calculator costs sixteen pounds. In a sale the shop reduces all prices by twenty five per cent. How much does the calculator cost in the sale?
12 On the answer sheet is part of a scale. What number is the arrow pointing to?
13 Imagine a hexagonal prism. How many of its faces are rectangles?
14 The distance between Jo's house and the shop is one kilometre. Jo is walking from her house to the shop. She is exactly three quarters of the way there. How many metres has she walked so far?
15 A football team scored three goals for every two they let in. They scored nine goals. How many did they let in?

"For the next set of questions you have fifteen seconds to work out each answer and write it down."

16 Which number is exactly half way between eighteen and thirty two?
17 Find the total of twenty four, thirty six and forty two.
18 A man was born in 1964. How old was he in the year two thousand?
19 One side of a square is six point two centimetres. What is the perimeter of the square?
20 Circle the expression shown on your answer sheet that has an answer nearest to fifty.

Test 6

Questions

"For this first set of questions you have five seconds to work out each answer and write it down."

1 Write the number four thousand and five in figures.
2 What is eighteen more than twenty?
3 What is three point two multiplied by ten?
4 One quarter of the children in a class are boys. What percentage are boys?
5 How many sides do three pentagons have in total?
6 What is twice thirty six?

"For the next set of questions you have ten seconds to work out each answer and write it down."

7 What is half of five point four?
8 How many seconds are there in one and a half minutes?
9 Add five and nineteen and then divide by eight.
10 A snail crawls thirty seven centimetres. How much further does it have to crawl to reach one metre?
11 One quarter of a number is four. What is the number?
12 On the answer sheet is part of a scale. What number is the arrow pointing to?
13 A shop has a sale. All items are twenty five per cent of their original prices. A video costs five pounds in the sale. What was its original price?
14 Dean walks six miles every day. How far does he walk in a week?
15 What is the probability that the day after June 6th will be June 7th?

"For the next set of questions you have fifteen seconds to work out each answer and write it down."

16 Look at the answer sheet. Draw a ring around the number that is nearest to four.
17 Twenty five per cent of a number is thirty. What is the number?
18 Which year is twenty three years before the year two thousand and two?
19 Look at the answer sheet. What is the size of angle A?
20 Which of the expressions on your answer sheet has an answer nearest to twenty?

"For this first set of questions you have five seconds to work out each answer and write it down."

1 Subtract nineteen from fifty one.
2 Two numbers have a total of one thousand. One of the numbers is eight hundred and fifty. What is the other number?
3 Write the number that is eight squared.
4 The time is quarter past three in the afternoon. Write this time as it would be shown on a twenty-four hour digital clock.
5 How many five pence pieces have the same value as five pounds?
6 The perimeter of a square is eighty eight centimetres. What is the length of one of its sides?

"For the next set of questions you have ten seconds to work out each answer and write it down."

7 Find the difference between four hundred and seventy and six hundred and ten.
8 Look at the number sequence on your answer sheet. What is the next number?
9 How much change from five pounds would you get if you spent exactly one pound twenty five pence?
10 If I add sixteen to a number I get thirty seven. What is the number?
11 A TV programme starts at twenty five past seven and lasts for forty minutes. What time does the programme end?
12 On the answer sheet is part of a scale. What number is the arrow pointing to?
13 How many teams of six can be made from a group of forty eight people?
14 Reading is one hundred and eighty kilometres from Bristol. Claire is driving from Reading to Bristol. After an hour she is three quarters of the way there. How many kilometres has she still to drive?
15 Circle the fraction that is equivalent to nought point seven.

"For the next set of questions you have fifteen seconds to work out each answer and write it down."

16 Add twenty six, twenty eight and nineteen.
17 Which number is exactly half way between fourteen and thirty?
18 A woman was born in 1953. How old was she in the year two thousand?
19 Circle the expression shown on your answer sheet that has an answer nearest to ten.
20 Look at the pie chart on the answer sheet. Two hundred people were asked what their favourite food was. How many people said their favourite food was chips?

Test 8

Questions

"For this first set of questions you have five seconds to work out each answer and write it down."

1 What is twenty five multiplied by five?
2 Write the number seventeen thousand, one hundred and four in figures.
3 Subtract thirty from one hundred and ten.
4 How many centimetres are there in one point six metres?
5 What number is seven less than three thousand?
6 What is three quarters as a percentage?

"For the next set of questions you have ten seconds to work out each answer and write it down."

7 An ice-cream costs ninety pence and a lolly costs seventy five pence. How much do they cost altogether?
8 The temperature yesterday was three degrees Celsius, and the temperature today is seven degrees colder. What is the temperature today?
9 Look at the answer sheet. In the equation, what is the value of x?
10 What is one third of six point nine?
11 Multiply eight by six.
12 Two numbers have a difference of twelve. One number is ninety one. What could the other number be?
13 Look at the answer sheet. Circle the number that is nearest to six kilograms.
14 Daniel slept for nine and a quarter hours without waking. He went to sleep at nine thirty p.m. What time did he wake?
15 Look at the answer sheet. What is the median of the numbers?

"For the next set of questions you have fifteen seconds to work out each answer and write it down."

16 Look at the answer sheet. Draw a ring around the largest of the five numbers.
17 A regular shape has a perimeter of forty five centimetres. Each side is nine centimetres. What is the name of the shape?
18 I have fifty pence. My friend has four times that much. How much do we have altogether?
19 Which number is exactly half way between one thousand nine hundred and two thousand?
20 Look at the answer sheet. What is the size of angle P?

"For this first set of questions you have five seconds to work out each answer and write it down."

1. Subtract forty five from ninety.
2. Two numbers have a total of two hundred. One of the numbers is eighty one. What is the other number?
3. Write the number that is seven squared.
4. The time is quarter past six in the evening. Write this time as it would be shown on a twenty-four hour digital clock.
5. How many fifty pence pieces have the same value as ten pounds?
6. The perimeter of a regular hexagon is sixty centimetres. What is the length of one of its sides?

"For the next set of questions you have ten seconds to work out each answer and write it down."

7. Find the difference between three hundred and ninety and five hundred and sixty.
8. Look at the number sequence on your answer sheet. What is the next number?
9. How much change from ten pounds would you get if you spent exactly eight pounds and fifteen pence?
10. A TV programme lasts for forty five minutes. It finishes at ten past five. What time does the programme start?
11. If I add fifteen to a number I get thirty two. What is the number?
12. Look at your answer sheet. What is the area of the rectangle?
13. How many teams of nine can be made from a group of fifty six people?
14. Calculate the total of three point six and four point nine.
15. Circle the decimal that is equivalent to sixty per cent.

"For the next set of questions you have fifteen seconds to work out each answer and write it down."

16. Seventy five per cent of a number is six. What is the number?
17. A square has an area of sixty four centimetres squared. What is the length of one of its sides?
18. A woman will have her thirty second birthday in the year two thousand and one. What year was the woman born?
19. Circle two fractions that are equivalent to one eighth.
20. Look at the pie chart on the answer sheet. Eighty people were asked what their favourite day of the week was. How many people said their favourite day was Saturday?

Test 10

Questions

"For this first set of questions you have five seconds to work out each answer and write it down."

1 What is the product of seven and three?
2 Write the number thirteen thousand and four in figures.
3 Subtract fifty from two hundred and twenty.
4 How many grams are there in one point nine kilograms?
5 Reduce two thousand by six.
6 What is thirty percent as a decimal?

"For the next set of questions you have ten seconds to work out each answer and write it down."

7 A number lies between forty and fifty. When it is divided by ten there is a remainder of four. What is the number?
8 The temperature yesterday was minus six degrees Celsius. The temperature today is three degrees warmer. What is the temperature today?
9 Look at the answer sheet. In the equation, what is the value of *x*?
10 What is one third of twelve point three?
11 Increase eighteen by thirty three.
12 Two numbers have a difference of nought point three. One number is six. What could the other number be?
13 Look at the answer sheet. Circle the number that is nearest to three quarters.
14 David slept for ten and a quarter hours without waking. He went to sleep at eight fifteen p.m. What time did he wake?
15 Look at the answer sheet. What is the median of the numbers?

"For the next set of questions you have fifteen seconds to work out each answer and write it down."

16 Look at the answer sheet. Draw a ring around the number that is a multiple of eight.
17 A regular shape has a perimeter of forty eight centimetres. Each side is six centimetres. What is the name of the shape?
18 I have forty pence. My friend has five times that much. How much do we have altogether?
19 Which number is exactly half way between two thousand six hundred and three thousand?
20 Look at the answer sheet. What is the size of angle G?

"For this first set of questions you have five seconds to work out each answer and write it down."

1 Add fifty eight and twenty one.
2 How many grams are there in three quarters of a kilogram?
3 Multiply fifteen by ten.
4 Write the decimal that is equivalent to seven tenths.
5 Divide twenty seven by nine.
6 How many twenty pence pieces have the same value as one pound forty pence?

"For the next set of questions you have ten seconds to work out each answer and write it down."

7 Add four point six and two point five.
8 Double four hundred and eighty.
9 A girl buys a forty pence magazine. She pays with a five pound note. How much change is she given?
10 A bag contains three red balls and one yellow ball. The probability of picking a red ball is three quarters. What is the probability of picking a yellow ball?
11 A coat costs twenty pounds. In a sale the shop reduces all prices by fifteen per cent. How much does the coat cost in the sale?
12 Look at the answer sheet. How many lines of reflective symmetry does the regular hexagon have?
13 Imagine a cube. How many vertices does it have?
14 A full bag of sugar weighs one kilogram. How many grams will it weigh when it is three quarters full?
15 In a class there are two boys to every girl. How many are girls if there are twenty two boys?

"For the next set of questions you have fifteen seconds to work out each answer and write it down."

16 One side of a square is six point two centimetres. What is the perimeter of the square?
17 Two numbers have a difference of two point two. The higher number is eleven point one. What is the lower number?
18 There are twice as many children as adults at a funfair. There are exactly one hundred and thirty two people at the fair. How many are adults?
19 Which year was thirteen years before the year two thousand and one?
20 Which number on your answer sheet is an approximate answer to the question 'What is seventeen per cent of forty nine'?

Test 12

Questions

"For this first set of questions you have five seconds to work out each answer and write it down."

1 What is the product of six and four?
2 How many days are there in eleven weeks?
3 Write the number ten thousand four hundred in figures.
4 Reduce five thousand by four.
5 How many centimetres are there in three point five metres?
6 Write seventy five percent as a fraction in its simplest form.

"For the next set of questions you have ten seconds to work out each answer and write it down."

7 A tin of beans costs thirty five pence and a loaf of bread costs seventy eight pence. How much do the beans and bread cost altogether?
8 The temperature yesterday was two degrees Celsius, and the temperature today is seven degrees colder. What is the temperature today?
9 What is one quarter of nine point six?
10 Look at the answer sheet. In the equation, what is the value of q?
11 A car park costs twenty five pence. Sixteen cars visit the car park. How much money do they pay in total?
12 The time is twenty past one in the afternoon. Write it as it would be shown on a twenty-four hour digital clock.
13 Look at the answer sheet. Circle the smallest of the five numbers.
14 Lucy slept for ten and a quarter hours without waking. She went to sleep at nine p.m. What time did she wake?
15 Look at the answer sheet. What is the modal value of the set of numbers?

"For the next set of questions you have fifteen seconds to work out each answer and write it down."

16 There are three times as many children as adults at a funfair. There are exactly one hundred and four people at the fair. How many are adults?
17 A regular shape has a perimeter of six point four centimetres. Each side is nought point eight centimetres. What is the name of the shape?
18 I have twenty two pence. My friend has four times that much. How much do we have altogether?
19 Which number is exactly half way between one thousand and four thousand?
20 Look at the answer sheet. What is the size of angle R?

5-second questions

1

2

3

4

5

6 cm

6

10-second questions

7

8 4, 7, 10, 13 …

9

10

11

12

13

14 km

15 $\frac{3}{4}$ $\frac{1}{3}$ $\frac{3}{10}$ $\frac{0}{3}$

9

15-second questions

16

17 16, 30

18 1913

19 3.5×1.5 35×1.5 0.15×35 0.35×15

20 people

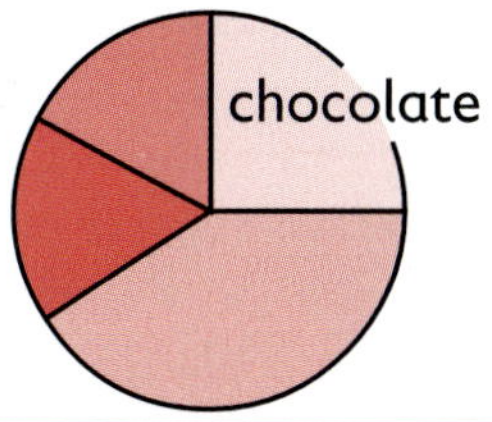

5

TOTAL

20

Test 4

Answer Sheet

5-second questions

1	
2	
3	
4	grams
5	
6	

6

10-second questions

7	seconds
8	
9	$n + 28 = 42$
10	
11	
12	
13	0.5 3.0 0.3 0.03 30.0
14	
15	3 6 4 5 3 8

9

15-second questions

16	5.7 6.1 6.09 6.11 5.8
17	centimetres
18	
19	25 12 121 100 51 49
20	bus station 12:52 shopping centre 12:58 swimming pool 13:11 cinema ....

5

TOTAL

20

5-second questions

1

2

3

4

5

6 metres

6

10-second questions

7

8

9 £ | 70p, £10

10

11 £ | £16

12

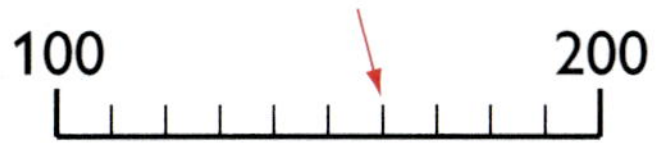

13

14 metres

15 | 3 : 2 9 goals

9

15-second questions

16 | 18, 32

17 | 24, 36, 42

18 | 1964

19 | 6.2 cm

20 $10.5 \div 2.3$ $105 \div 2.3$ $1.05 \div 2.3$ $1.05 \div 23$

5

TOTAL

20

Test 6
Answer Sheet

5-second questions

1

2

3

4 %

5

6

6

10-second questions

7

8 seconds

9

10 centimetres

11

12

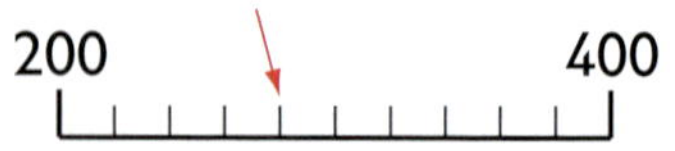

13 £

14 miles

15

9

15-second questions

16 4.5 3.4 4.16 3.8 4.44

17

18

19 A =

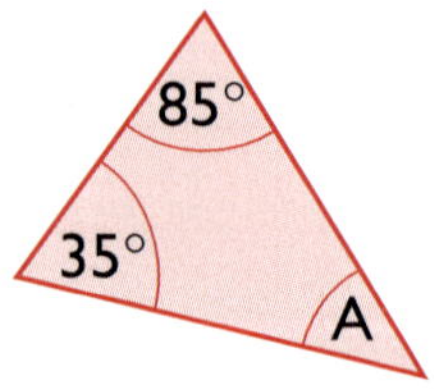

20 4.2×5.5 6.1×4.8 1.9×5.2 3.9×4.9

5

TOTAL

20

5-second questions

1

2

3

4

5

6 cm

6

10-second questions

7

8 4, 8, 16, 32, …

9 £

10

11

12

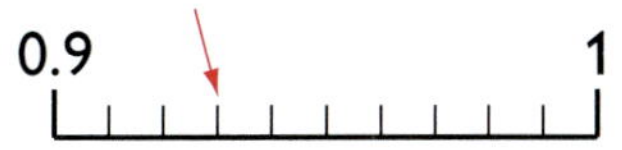

13

14 km

15 $\frac{0}{7}$ $\frac{1}{7}$ $\frac{7}{10}$ $\frac{7}{1}$

9

15-second questions

16

17 14, 30

18 1953

19 3.1×3.1 2.5×2.5 0.15×0.15 4.12×4.12

20

people

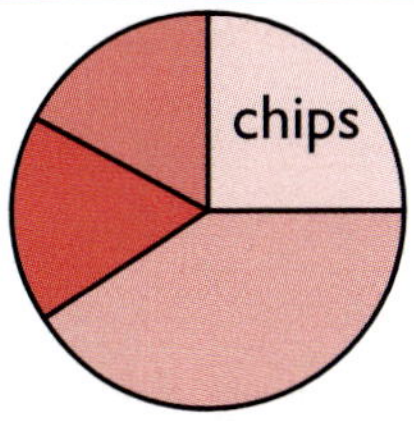

5

TOTAL

20

Test 8
Answer Sheet

5-second questions

1

2

3

4 cm

5

6 %

6

10-second questions

7 £

8 °C

9 | $x - 34 = 66$

10 | 6.9

11

12 | 91

13 6.4 kg 610 g 6.5 kg 590 g 6300 g

14

15 | 3 5 5 7 8 9

9

15-second questions

16 0.74 0.06 0.9 0.51 0.2

17

18 £

19

20 P = °

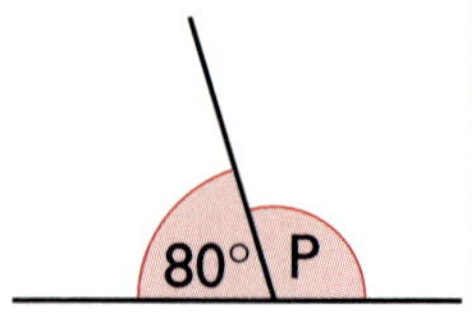

5

TOTAL

20

5-second questions

1	
2	grams
3	
4	
5	
6	

6

10-second questions

7		
8		
9	£	40p, £5
10		
11	£	£20
12		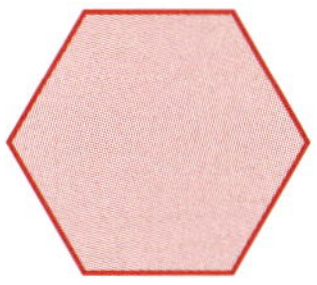
13		
14		grams
15		2 : 1 22 boys

9

15-second questions

16		6.2 cm
17		11.1
18		132
19		2001
20		17 34 68 $8\frac{1}{2}$ 49 $24\frac{1}{2}$

5

TOTAL

20

Test 12

Answer Sheet

5-second questions

1	
2	days
3	
4	
5	cm
6	

6

10-second questions

7	£	
8	°C	
9		9.6
10		$37 + q = 63$
11		25p, 16
12		
13	6.45 0.54 0.6 6.4 0.64	
14		
15		4 6 3 2 3 5 3

9

15-second questions

16		104
17		6.4 cm
18	£	
19		
20	R = °	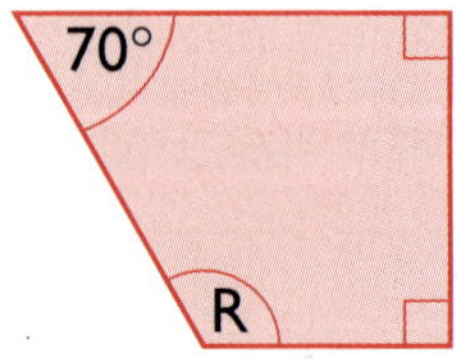

5

TOTAL

20

Test 1

1 35
2 120
3 9018
4 2600 g
5 0.75
6 1997
7 £2.79
8 −3°C
9 34
10 2.1
11 200
12 48 or 12
13 640 cm
14 6.30 a.m.
15 $\frac{6}{10}$
16 4.7
17 hexagon
18 £2.40
19 1750
20 100°

Notes:

- The mean value is found by finding the total of the scores and dividing by the number of scores, in this case 3.
- To find a number half way between two numbers, add them and divide by two.
- The angles inside a four-sided shape (a quadrilateral) total 360°.

Test 2

1 40011
2 39
3 3.2
4 12
5 25
6 72
7 4.1
8 6
9 28 cm
10 7.15
11 £8.50
12 0.17
13 4
14 250 ml
15 8
16 35
17 101
18 1972
19 26 cm
20 27

Test 3

1 34
2 350
3 25
4 13:45
5 50
6 20 cm
7 280
8 16
9 £4.25
10 13
11 6.10
12 0.53
13 8
14 360 km
15 $\frac{3}{10}$
16 48
17 23
18 1962
19 35 × 1.5
20 40

Notes:

- The sequence in Q8 is increasing by three each time.

Test 4

1 63
2 77
3 10004
4 250 g
5 $\frac{1}{4}$
6 1200
7 600
8 100
9 14
10 36
11 250
12 18:40
13 0.3
14 9.45
15 3
16 6.09
17 20 cm
18 hemispheres
19 25, 121, 100, 49
20 13:15

Notes:

- The mode is the number that occurs most often.
- Square numbers include 1, 4, 9, 16, 25, 36, 49, 64, 81, 100, 121

Answers – Mental Tests

Test 5

1 30016
2 37
3 150
4 9
5 20
6 1500
7 540
8 10.1
9 £9.30
10 ten to eight or 7.50
11 £12
12 160
13 6
14 750 m
15 6
16 25
17 102
18 36
19 24.8 cm
20 105 ÷ 2.3

Notes:

- To find a number half way between two numbers, add them and divide by two.

Test 6

1 4005
2 38
3 32
4 25%
5 15
6 72
7 2.7
8 90
9 3
10 63 cm
11 16
12 280
13 £20
14 42 miles
15 1 or certain
16 4.16
17 120
18 1979
19 60°
20 3.9 × 4.9

Notes:

- The probability of a definite occurrence is described as certain and has a probability of 1. (Impossible situations have a probability of zero.)
- The angles inside a triangle have a total of 180°.

Test 7

1 32
2 150
3 64
4 15:15
5 100
6 22 cm
7 140
8 64
9 £3.75
10 21
11 8.05 or five past eight
12 0.93
13 8
14 45 km
15 $\frac{7}{10}$
16 73
17 22
18 47
19 3.1 × 3.1
20 50

Notes:

- To find a number half way between two numbers, add them and divide by two.

Test 8

1 125
2 17104
3 80
4 160 cm
5 2993
6 75%
7 £1.65
8 −4°C
9 100
10 2.3
11 48
12 103 or 79
13 6300 g
14 6.45 a.m.
15 6
16 0.9
17 pentagon
18 £2.50
19 1950
20 100°

Notes:

- The median value is the number in the middle when the set is arranged in order. If two numbers occur in the middle, the median is exactly half way between the two numbers.

Test 9

1 45
2 119
3 49
4 18:15
5 20
6 10 cm
7 170
8 25
9 £1.85
10 4.25
11 17
12 $32\ cm^2$
13 6
14 8.5
15 0.6
16 8
17 8 cm
18 1969
19 $\frac{4}{32}$, $\frac{8}{64}$
20 60

Notes:

- Children often confuse area and perimeter. Perimeter is the distance all the way around the edge of a shape. The area of a shape is the amount of surface inside the shape. For rectangles, this can be calculated by multiplying the length by the breadth.

Test 10

1 21
2 13004
3 170
4 1900 g
5 1994
6 0.3
7 44
8 −3°C
9 8 or 8.0
10 4.1
11 51
12 6.3 or 5.7
13 0.8
14 6.30 a.m.
15 5
16 728
17 octagon
18 £2.40
19 2800
20 160°

Notes:

- A regular shape has equal sides and equal angles.

Test 11

1 79
2 750 g
3 150
4 0.7
5 3
6 7
7 7.1
8 960
9 £4.60
10 one quarter
11 £17
12 6
13 8
14 750 g
15 11
16 24.8 cm
17 8.9
18 44
19 1988
20 $8\frac{1}{2}$

Notes:

- A vertex (plural: vertices) is the mathematical word for the corner of a shape.

Test 12

1 24
2 77 days
3 10400
4 4996
5 350 cm
6 $\frac{3}{4}$
7 £1.13
8 −5°C
9 2.4
10 26
11 £4
12 13:20
13 0.54
14 7.15 a.m.
15 3
16 26
17 octagon
18 £1.10
19 2500
20 110°

Notes:

- The product of two numbers is found by multiplying them together.
- The angles inside a four sided shape (a quadrilateral) total 360°.

National Curriculum Levels

The Key Stage 2 National Tests are levelled according to the child's total score for the written Tests A and B and one mental test, similar to those in this book.

Mark	24 or below	25–51	52–79	80–100
Level	Level 1/2	Level 3	Level 4	Level 5

Scores for the mental test can be broadly broken down as follows:

Mark	0–4	5–10	11–15	16–20
Level	Level 1/2	Level 3	Level 4	Level 5

For practice in the written tests, look for the Key Stage 2 Maths books in this series. If your child needs more practice in any Maths topics, use the WH Smith Key Stage 2 Maths Revision Guide.

WRITING NUMBERS IN FIGURES

Write in figures the number:

1 two thousand and six

2 four thousand and three

3 five thousand and eleven

4 six thousand and fifteen

5 ten thousand and four

6 ten thousand and eighteen

7 twenty thousand and seven

8 thirty thousand and nineteen

9 forty thousand and twelve

10 fifty thousand and two

DOUBLING AND HALVING

11 What is **double**:

seventeen? forty two? fifty five?

12 What is **twice**:

thirty eight? sixty three? one point four?

13 What is **half**:

fifty two? thirty six? eighty six?

14 What is **half**:

sixty eight? thirty four? seventy two?

MORE THAN/LESS THAN

What is:

15 sixteen more than twenty?

16 eighteen more than fifteen?

17 thirteen less than fifty?

18 thirty six less than fifty?

19 three less than two thousand?

20 two less than five hundred?

ADDING AND SUBTRACTING

21 Two numbers total one thousand. One of the numbers is four hundred and fifty. What is the other number?

22 Two numbers total one thousand. One of the numbers is eight hundred and twenty five. What is the other number?

Find the total of:

23 thirteen and twelve

sixteen and seventeen

24 nineteen and eleven

fourteen and sixteen

Subtract:

25 sixteen from seventy

26 sixty from one hundred and ten

27 nineteen from fifty one

28 forty from two hundred and twenty

MULTIPLYING AND DIVIDING

What is:

29 eight multiplied by five?

30 nine multiplied by four?

31 six multiplied by twenty?

32 four point two multiplied by ten?

33 seven point one multiplied by ten?

34 thirty six divided by six?

35 twenty seven divided by nine?

36 fifty six divided by eight?

37 thirty four divided by ten?

38 one hundred and thirty divided by ten?

39 Write the number that is 6 squared.

40 Write the number that is 7 squared.

41 Write the number that is 9 squared.

42 How many sides do four pentagons have in total?

43 How many sides do six hexagons have in total?

44 How many sides do seven octagons have in total?

FRACTIONS, DECIMALS AND PERCENTAGES

45 Three quarters of the children in a class are girls. What percentage are girls?

46 Four tenths of the children in a class are boys. What percentage are boys?

47 One quarter of the children in a class are boys. What percentage are boys?

What is:

48 one third of eighteen?

49 two thirds of twenty seven?

50 one quarter of twenty?

51 one fifth of fifty?

52 one eighth of sixty four?

53 three quarters of twelve?

54 What is one quarter as a decimal?

55 What is seven tenths as a decimal?

56 What is two hundredths as a decimal?

57 What is nought point five as a fraction?

58 What is nought point one as a fraction?

59 What is nought point nine as a fraction?

PROBABILITY

60 A bag contains four green cubes and one white cube. The probability of picking a green cube is four fifths. What is the probability of picking a white cube?

61 One blue cube and five white cubes are in a bag. The probability of picking a blue cube is one sixth. What is the probability of picking a white cube?

MEASUREMENT AND MONEY

How many:

62 millilitres are there in three and a half litres?

63 grams are there in two point six kilograms?

64 grams are there in a quarter of a kilogram?

65 metres are there in one and a half kilometres?

66 ten pence coins have the same value as three pounds?

67 two pence pieces have the same value as one pound?

68 five pence coins have the same value as one pound?

69 five pence pieces have the same value as five pounds?

70 months are there in ten years?

71 days are there in eleven weeks?

72 hours are there in three days?

MEASUREMENT AND MONEY

73 The time is quarter to four in the afternoon. Write this time as it would be shown on a twenty-four hour digital clock.

74 The time is quarter past two in the afternoon. Write this time as it would be shown on a twenty-four hour digital clock.

75 The perimeter of a square is 12 cm. What is the length of one of its sides?

76 The perimeter of a square is 8 cm. What is the length of one of its sides?

77 The perimeter of a regular pentagon is 50 cm. What is the length of one of its sides?

78 The perimeter of a regular hexagon is 24 cm. What is the length of one of its sides?

CALCULATIONS

1 Add five and eleven and then divide by four.

2 Multiply seven by four and then subtract nine.

3 Add twelve and nine and then divide by three.

4 Divide one hundred by four and then subtract seven.

5 Emma walks four miles every day. How far does she walk in a week?

6 Sam cycles nine miles every day. How far does he cycle in a week?

7 James cycles six miles a week. In how many weeks will he have cycled thirty six miles?

8 Paul cycles eight miles a week. In how many weeks will he have cycled sixty four miles?

9 Thirty six eggs in boxes of six. How many boxes?

10 How many eight centimetre lengths of ribbon can be cut from a length of forty eight centimetres?

11 How many six pence chews can I buy with fifty four pence?

READING SCALES

What numbers are the arrows pointing to?

12 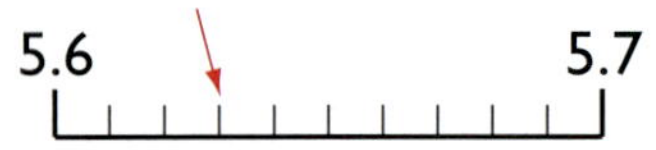

13 

NUMBER PATTERNS AND ALGEBRA

What is the next number in the sequence?

14 16, 21, 26, 31, ☐   39, 32, 25, 18, ☐

15 1, 4, 9, 16, ☐   43, 32, 21, 10, ☐

In the equation, what is the value of x?

16 $x + 15 = 31$ ☐ $x - 20 = 57$ ☐

17 $27 + x = 42$ ☐ $51 - x = 39$ ☐

FRACTIONS, DECIMALS AND PERCENTAGES

18 One quarter of a number is six. What is the number? ☐

19 One third of a number is eight. What is the number? ☐

20 One fifth of a number is ten. What is the number? ☐

21 One ninth of a number is eight. What is the number? ☐

22 One sixth of a number is three. What is the number? ☐

FRACTIONS, DECIMALS AND PERCENTAGES

23 A shop has a sale. Ten per cent is taken off all prices.
What is the sale price of a book that was twenty pounds? ☐

24 In a sale, items cost twenty five per cent of their original prices.
A video costs £6 in the sale. What was its original price? ☐

25 In a sale, items cost seventy five per cent of their original prices.
A video costs £30 in the sale. What was its original price? ☐

26 Circle two equivalent fractions in each row.

$\frac{1}{2}$	$\frac{2}{3}$	$\frac{3}{4}$	$\frac{1}{4}$	$\frac{4}{6}$	$\frac{7}{8}$	$\frac{7}{10}$
$\frac{1}{3}$	$\frac{5}{6}$	$\frac{7}{9}$	$\frac{1}{4}$	$\frac{5}{6}$	$\frac{2}{8}$	$\frac{1}{2}$
$\frac{1}{10}$	$\frac{2}{3}$	$\frac{4}{12}$	$\frac{1}{6}$	$\frac{2}{5}$	$\frac{4}{8}$	$\frac{1}{3}$

27 What is 10% of:

280? ☐ 470? ☐ 30? ☐ 1000? ☐

28 What is 25% of:

28? ☐ 48? ☐ 60? ☐ 1000? ☐

29 What is 75% of:

28? ☐ 48? ☐ 60? ☐ 1000? ☐

MEASUREMENT AND MONEY

30 Draw a ring around the approximate height of a door.

200 m   20 m   20 cm   2000 cm   200 cm

31 Draw a ring around the approximate mass of a bag of sugar.

100 g   10 kg   100 kg   1000 g   10 g

32 A snail crawls sixty three centimetres. How much further does it have to crawl to reach one metre?

33 A snail crawls eighty one centimetres. How much further does it have to crawl to reach one metre?

34 What is two hundred and seventy one centimetres to the nearest metre?

35 What is two thousand one hundred grams to the nearest kilogram?

36 I start with a five pound note. I spend one pound forty pence. How much have I now?

37 A programme started at twenty past seven. It finished at five past eight. How many minutes did the programme last?

38 A film started at quarter to nine. It finished at ten o'clock. How many minutes did the film last?

FRACTIONS, DECIMALS AND PERCENTAGES

1 Draw a ring around the number that is nearest to four.

3.2 4.46 4.9 3.19 3.7

2 Draw a ring around the number that is nearest to six.

6.11 6.46 6.099 5.8 5.59

3 Twenty five per cent of a number is sixteen.
What is the number?

4 Seventy five per cent of a number is sixty three.
What is the number?

5 Twenty five per cent of a number is eighteen.
What is the number?

6 Thirty per cent of a number is sixty.
What is the number?

7 One fifth of a number is fifteen.
What is the number?

CALCULATIONS OF TIME

8 Which year is sixteen years before the year two thousand and two?

9 Which year is nineteen years before the year two thousand and eight?

10 A man was born in 1893. What year was the man's ninety ninth birthday?

NUMBER KNOWLEDGE

11 Which number is exactly half way between one thousand five hundred and two thousand?

12 Which number is exactly half way between twenty two and fifty?

13 Which number is exactly half way between fourteen and thirty?

14 Draw a ring around **three** numbers that are multiples of twenty.

120 580 330 500 170

15 Draw a ring around **three** numbers that are factors of thirty.

36 60 1 6 15

16 Draw a ring around **three** numbers that are prime numbers.

17 36 51 71 19

MONEY PROBLEMS

17 I went to the shop with five pounds. I spent forty two pence and then fifty seven pence. How much did I have left?

18 I went to the shop with ten pounds. I spent seventy pence and then one pound and fifteen pence. How much did I have left?

19 I have eighty pence. My friend has three times that much. How much do we have altogether?

SHAPE AND MEASUREMENT

20 What is the size of angle C?

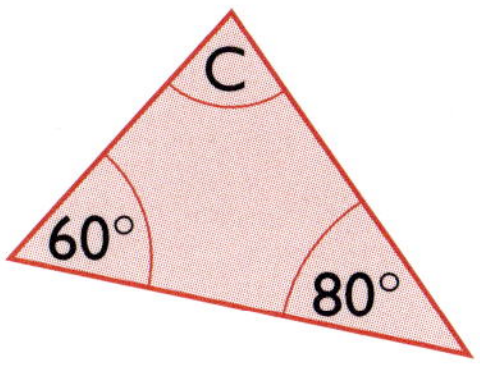

21 What is the size of angle A?

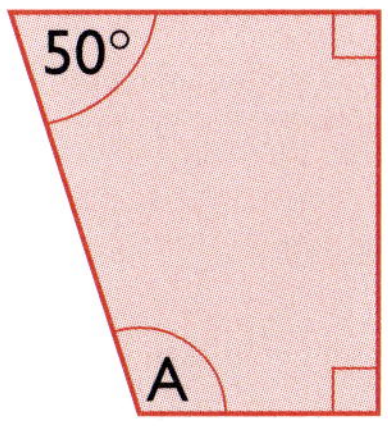

22 What is the size of angle B?

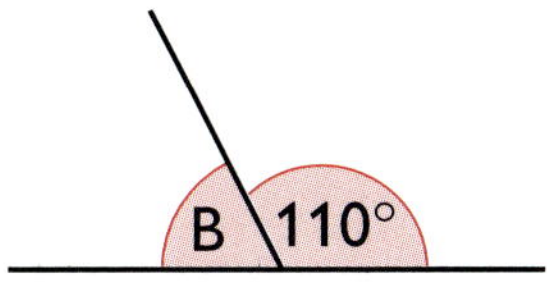

23 A regular shape has a perimeter of 48 cm. Each side is 6 cm. What is the name of the shape?

24 Each of the sides of a pentagon is four point two centimetres. What is its perimeter?

DATA HANDLING

25 Look at the pie chart. One hundred and sixty people were asked what their favourite food was. How many people said their favourite food was crisps?

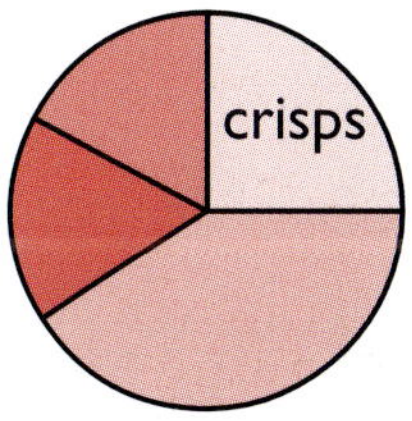

26 Here are some start and finish times of TV programmes. How long did the documentary last?

	Start	Finish
Documentary	14:25	15:20
Travel	16:33	17:06
Wildlife	19:36	20:27

Answers

Practice Pages

5-second questions

1 2006
2 4003
3 5011
4 6015
5 10004
6 10018
7 20007
8 30019
9 40012
10 50002

11 34, 84, 110
12 76, 126, 2.8
13 26, 18, 43
14 34, 17, 36

15 36
16 33
17 37
18 14
19 1997
20 498

21 550
22 175
23 25, 33
24 30, 30
25 54
26 50
27 32
28 180

29 40
30 36
31 120
32 42
33 71
34 6
35 3
36 7
37 3.4
38 13
39 36
40 49
41 81
42 20
43 36
44 56

45 75%
46 40%
47 25%
48 6
49 18
50 5
51 10
52 8
53 9
54 0.25
55 0.7
56 0.02
57 $\frac{1}{2}$
58 $\frac{1}{10}$
59 $\frac{9}{10}$

60 $\frac{1}{5}$
61 $\frac{5}{6}$

62 3500 ml
63 2600 g
64 250 g
65 1500 m
66 30
67 50
68 20
69 100
70 120
71 77
72 72

73 15:45
74 14:15
75 3 cm
76 2 cm
77 10 cm
78 4 cm

10-second questions

1 4
2 19
3 7
4 18
5 28
6 63
7 6
8 8
9 6
10 6
11 9

12 5.63, 280
13 540, 0.91

14 36, 11
15 25, –1
16 16, 77
17 15, 12

18 24
19 24
20 50
21 72
22 18

23 £18
24 £24
25 £40
26 $\frac{2}{3}$, $\frac{4}{6}$
$\frac{1}{4}$, $\frac{2}{8}$
$\frac{4}{12}$, $\frac{1}{3}$
27 28, 47, 3, 100
28 7, 12, 15, 250
29 21, 36, 45, 750

30 200 cm
31 1000 g
32 37 cm
33 19 cm
34 3 m
35 2 kg
36 £3.60
37 45 mins
38 75 mins

15-second questions

1 3.7
2 6.099
3 64
4 84
5 72
6 200
7 75

8 1986
9 1989
10 1992

11 1750
12 36
13 22
14 120, 580, 500
15 1, 6, 15
16 17, 71, 19

17 £4.01
18 £8.15
19 £3.20

20 40°
21 130°
22 70°
23 octagon
24 21 cm

25 40
26 55 mins